Gaetano Lanza

Some Tests of the Strength of Cast Iron

made in the Laboratory of applied mechanics of the Massachusetts

institute of technology

Gaetano Lanza

Some Tests of the Strength of Cast Iron
made in the Laboratory of applied mechanics of the Massachusetts institute of technology

ISBN/EAN: 9783337182168

Printed in Europe, USA, Canada, Australia, Japan

Cover: Foto ©berggeist007 / pixelio.de

More available books at **www.hansebooks.com**

ᴐᴛᴇ.—This paper is sent to you that you may examine it in advance of the meeting, and prepare any discussion of it which you may wish to present.

It is issued to the membership in confidence, and with the distinct understanding that it is not to be given to the press or to the public until after it has been presented at the meeting.

As there will be no supply of extra copies there, and papers are liable to be read by abstract only, preserve this copy for your use, and

BRING THIS COPY WITH YOU TO THE MEETING.

(Subject to Revision.)

Presented at the XVIIIth Meeting, Scranton, 1888.
American Society of Mechanical Engineers.
Advance sheets from Vol. X. Transactions.

CCCXX.

SOME TESTS OF THE STRENGTH OF CAST IRON MADE IN THE LABORATORY OF APPLIED MECHANICS OF THE MASSACHUSETTS INSTITUTE OF TECHNOLOGY.

BY GAETANO LANZA
(Member of the Society.)

WITH HEYWOOD COCHRAN, JOHN K. BURGESS, MAURICE A. VIELÉ, HENRY F. EASTMAN, AND WM. H. GERRISH.

The object of this paper is to give a brief account of several sets of tests upon the strength and other resisting properties of cast iron, carried on in the laboratory of Applied Mechanics of the Massachusetts Institute of Technology, of which the results are, it is believed, of sufficient practical value to render them worthy of record.

The experiments referred to have formed the subjects of three graduating thesis, viz.:

1st. An investigation upon the modulus of elasticity and some other properties of cast iron, by Heywood Cochran of the class of 1885.

2d. An investigation of the tensile and the transverse strengths of cast iron, and a comparison of their respective moduli of elasticity, by John K. Burgess and Maurice A. Vielé, of the class of 1886.

3d. Experiments upon pulleys, keys and set screws, by Henry F. Eastman and William H. Gerrish, of the class of 1888.

The first portion of the work relates especially to the modulus of elasticity, and the limit of elasticity of common cast iron, both planed and unplaned, and of gun iron.

The main portion of the experiments, however, are upon the transverse strength of cast iron when used in the forms of window lintels and of pulleys.

The reason for undertaking these tests was, that it is well known that the modulus of rupture of cast iron varies greatly, according to the form of the casting, and the manner of using it; and it was considered desirable to obtain some experimental results which should be applicable to the forms mentioned.

Some experiments were also made upon the strength of keys of cast iron, wrought iron, and steel, and upon the holding power of set screws, all of which are recorded here.

SUMMARY OF THE FIRST SET OF EXPERIMENTS—BY MR. HEYWOOD COCHRAN.

The object of the thesis was to determine the values of the modulus of elasticity, and of the limit of elasticity of certain kinds of common cast iron, and of gun iron, and of re-testing the specimens.

The common iron consisted of a half-and-half mixture of Lake Superior magnetic and Harrington irons, the last being made from an English bog ore.

The gun iron consisted of a half-and-half mixture of Muirkirk, Md., and remelted Salisbury irons.

The chemical analyses as far as determined were as follows:

	Gun Iron.	Common Iron.
	%	%
Total carbon	3.51	——
Graphite	2.80	——
Sulphur	0.133	0.173
Phosphorus	0.155	0.413
Silicon	1.140	1.89

The test specimens, all of which were cast at the South Boston Iron Foundry, were twenty-six inches long and square in section; those tested with the skin on being very nearly one inch square, and those tested with the skin removed being cast nearly one and one-quarter inches square, and afterwards planed down to one inch square.

All were of the same section throughout their entire length.

The tables of tests will now be given:

TEST NO. 1.

UNPLANED COMMON IRON.

Gauged length, 13".3125.

Area of section, 1.0455 square inches.

Loads Applied.	Elongations, Inches.	Sets, Inches.	E
500	0.0000		
1,000	0.0004		18,148,370
1,500	0.0008		16,977,500
2,000	0.0012		16,608,450
500		0.0000	
2,500	0.0017		15,433,500
3,000	0.0023		14,148,000
3,500	0.0028		13,890,720
500		0.0002	
4,000	0.0032		13,926,800
4,500	0.0036		14,148,000
5,000	0.0042		13,807,000
500		0.0004	
5,500	0.0048		13,344,800
500		0.0004	
6,000	0.0052		13,533,540
500		0.0004	
6,500	0.0057		13,521,900
6,500	0.0056		
500		0.0004	
7,000	0.0061		13,568,140
500		0.0006	
7,500	0.0066		13,504,800
500		0.0008	

Tensile strength, 23,000 lbs. per sq. inch.

With a load of 11,000 lbs. the piece broke unexpectedly in the upper clamps, due to the fact that these clamps did not bind the piece as they should have done, but rather pinched it at its lower end. Then, too, the load was very suddenly applied. Upon re-testing, the piece broke with a load of 24,000 lbs., or 23,000 lbs. per square inch. A load of 6,500 lbs. was left on for seventeen hours and a half without producing any additional elongation. The position of the fracture was just outside the upper clamps.

TEST NO. 2.

UNPLANED COMMON IRON.

Gauged length, 13″.5938.

Area of section, 1.0754 square inches.

Loads Applied.	Elongation, Inches.	Sets, Inches.	E.
500	0.0000		
1,000	0.0004		18,058,140
1,500	0.0007		18,058,140
2,000	0.0011		17,237,310
2,500	0.0015		16,854,260
3,000	0.0020		16,206,000
3,500	0.0025		15,168,840
4,000	0.0030		14,997,430
4,500	0.0034		15,093,361
5,000	0.0039		14,585,430
500		0.0000	
5,500	0.0046		13,789,900
6,000	0.0050		13,904,800
6,500	0.0057		13,806,000
500		0 0003	
7,000	0.0063		13,146,330
7,500	0.0069		12,917,500
500		0.0005	
8,000	0.0075		12,640,700
500		0.0006	
8,500	0.0082		12,408,040
500		0.0005	
9,000	0.0095		
500		0.0009	

Tensile strength, 23,000 lbs. per square inch.

TEST NO. 3.

UNPLANED COMMON IRON.

Gauged length, 13″.4883.

Area of section, 1.0614 square inches.

Loads Applied.	Elongation, Inches.	Sets, Inches.	E.
500	0.0000		
1,000	0.0004		18,154,300
1,500	0.0008		16,398,300
2,000	0.0012		15,561,000
2,500	0.0017		14,950,600
3,000	0.0022		14,607,000
3,500	0.0026		14,523,460
4,000	0.0031		14,347,800
4,500	0.0037		13,926,600
5,000	0.0042		13,615,800
500		0.0001	
5,500	0.0050		12,771,900

At the end of the test a load of 9,000 lbs. was left upon the piece for seventy hours.

TEST NO. 4.

SAME SPECIMEN RE-TESTED.

Loads Applied.	Elongation, Inches.	Sets, Inches.	E.
500	0.0000		
1,000	0.0003		19,550,800
1,500	0.0007		18,154,300
2,000	0.0011		17,732,140
2,500	0.0015		17,528,300
3,000	0.0019		16,721,100
3,500	0.0023		16,575,700
4,000	0.0029		15,606,350
4,500	0.0035		14,734,000
5,000	0.0040		14,386,450
5,500	0.0048		13,237,530
500		0.0002	
6,000	0.0052		13,376,850
6,500	0.0058		13,146,240
500		0.0002	
7,000	0.0062		13,322,900
7,500	0.0066		13,478,200
500		0.0001	
8,000	0.0071		13,414,000
8,500	0.0076		13,376,900
500		0.0001	
9,000	0.0081		13,335,600
9,500	0.0086		13,299,100
500		0.0001	
10,000	0.0093		13,021,470
500		0.0002	
10,500	0.0100		12,771,900
500		0.0003	

The load of 5,500 lbs. was left upon the piece for two hours. At the end of this test a load of 12,000 lbs. was left upon the piece, this being above the limit of elasticity.

TEST NO. 5.

SAME PIECE RE-TESTED A SECOND TIME.

Loads Applied.	Elongation, Inches.	Sets, Inches.	E.
500	0.0000		
1,000	0.0004		14,950,630
1,500	0.0009		14,950,630
2,000	0.0013		14,950,630
2,500	0.0017		14,734,000
3.000	0.0022		14,606,900
3,5C0	0.0027		14,386,450
4,000	0.0032		14,120,030
4,500	0.C037		13,926,640
5,000	0.0042		13,779,800
5,500	0.0047		13,664,540
6,000	0.0052		13,571,700
6,500	0.0057		13,376,900
7,000	0.0062		13,322,900
7,500	0.0067		13,227,670
8,000	0.0073		13,146,230
8,500	0.0078		12,992,535
9,000	0.0084		12,859,320
9,500	0.0090		12,708,000
10,000	0.0095		12,708,000
10,500	0.0101		12,613,430
11,000	0.0107		12,499,700
500		0.0003	
11,500	0.0111		12,593 540
12,000	0.0117		12,490,800
12,500	0.0124		12,298,000
500		0.0007	
13,000	0.0130		12,266,430
500		0.0009	

Tensile strength, 20,200 lbs. per square inch.

TEST NO. 6.

UNPLANED GÚN IRON.

Gauged Length, 13."5625
Area of section,   1.0506 sq. inch.

Loads applied.	Elongation, inches.	Sets, inches.	E
500	0.0000		
1,000	0.0003		21,505,580
1,500	0.0007		18,441,850
2,000	0.0011		18,441,850
2,500	0.0015		17,805,920
3,000	0.0019		17,445,000
3,500	0.0023		17,212,400
4,000	0.0027		17,050,000
4,500	0.0031		16,930,200
5,000	0.0036		16,363,900
500	0.	0.0001	
5,500	0.0041		15,937,400
6,000	0.0046		15,435,000
6,500	0.0050		15,491,150
7,000	0.0055		15,256,440
500		0.0005	
7,500	0.0060		15,187,400
500		0.0005	
8,000	0.0063		15,368,480
500		0.0005	
8,500	0.0068		15,192,300
9,000	0.0074		14,929,830

Tensile strength, 27,000 lbs. per sq. in.

The piece broke first with a load of 18,900 lbs., exhibiting a bad flaw, and then, upon being re-tested broke at 28,450 lbs., or about 27,000 lbs. per sq. inch, as given above.

TEST NO. 7.

UNPLANED GUN IRON.

Gauged length, 13."3906.
Area of section, 1.0630 sq. in.

Loads applied.	Elongation, inches.	Sets, inches.	E.
500	0.0000		
1,000	0.0003		22,903,620
1,500	0.0005		23.994,800
2,000	0.0008		22,903,620
2,500	0.0012		21,907,800
3,000	0.0016		20,316,700
3,500	0.0020		19,380,000
4,000	0.0024		18,761,470
4,500	0.0027		18,662,200
5,000	0.0031		18,286,000
5,500	0.0035		18,125,160
6,000	0.0039		17,995,300
6,500	0.0043		17,680,000
7,000	0.0047		17,421,370
7,500	0.0051		17,205,540
8,000	0.0055		17,177,700
500		0.0000	
8,500	0.0059		17,080,670
9,000	0.0064		16,862,100
500		0.0000	
9,500	0.0068		16,672,500
10,000	0.0072		16,621,370
500		0.0001	
10,500	0.0078		16,254,180
11,000	0.0082		16,130,300
500		0.0003	
11,500	0.0087		16,019,300
500		0.0003	
12,000	0.0092		15,746,240
500		0.0005	
12,500	0.0098		15,504,000
500		0.0007	
13,000	0.0135		15,213,760
500		0.0008	

A load of 13,250 lbs. remained upon this piece for seventeen hours, this load being just above the elastic limit.

TEST NO. 8.

SAME SPECIMEN RE-TESTED.

Loads applied.	Elongation, inches.	Sets, inches.	E.
500	0.0000		
1,000	0.0003		20,995,000
1,500	0.0006		20,995,000
2,000	0.0009		20,427,550
2,500	0.0013		19,760,000
8,000	0.0017		19,076.300
3,500	0.0021		18,212,500
4,000	0.0025		17,995,700
4,500	0.0028		17,836,440
5,000	0.0032		17,714,500
5,500	0.0036		17,495,820
6,000	0.0040		17,429,800
6,500	0.0044		17,275,870
7,000	0.0048		17,147,750
7,500	0.0052		16,876,350
8,000	0.0056		16,796,000
500		0.0000	
8,500	0.0061		16,588,250
9,000	0.0065		16,536,580
9,500	0.0069		16,490,600
10,000	0.0073		16,449,700
500		0.0002	
10,500	0.0077		16,306,800
11,000	0.0081		16,279,200
500		0.0002	
11,500	0.0085		16,254,180
12,000	0.0089		16,231.410
12,500	0.0093		16,210,600
13,000	0.0098		16,108,680
500		0.0002	
13,500	0.0102		16,094,430
14,000	0.0108		15,782,750
14,500	0.0112		15,711,200
500		0.0003	
15,000	0.0118		15,446,600
500		0.0005	
16,000	0.0130		

A load of 16,000 lbs. was left upon the piece for 22 hours.

TEST NO. 9.

SAME SPECIMEN RE-TESTED A SECOND TIME.

Loads Applied.	Elongation, Inches.	Sets, Inches.	E
500	0.0000		
1,000	0.0003		19,880,000
1,500	0.0006		20,156,200
2,000	0.0009		20,427,550
2,500	0.0012		20,995,000
3,000	0.0016		20,317,700
3,500	0.0019		19,681,700
4,000	0.0023		19,169,300
4,500	0.0028		18,323,000
5,000	0.0032		17,995,700
5,500	0.0035		17,868,100
6,000	0.0039		17,995,700
6,500	0.0043		17,784,000
7,000	0.0047		17,703,890
7,500	0.0050		17,548,000
8,000	0.0054		17,495,820
8,500	0.0058		17.450,400
9,000	0.0062		17,420,450
9,500	0.0065		17,375,160
10,000	0.0071		16,974,670
10,500	0.0075		16,796,000
11,000	0.0080		16.637,530
11,500	0.0084		16,545,300
500		0.0000	16,415.340
12,000	0.0088		
12,500	0.0093		16,342,042
13,000	0.0097		16,233,240
13,500	0.0102		16,134,100
14,000	0.0106		16,043,300
500		0.0002	
14,500	0.0111		15,960,000
15,000	0.0115		15.883,000
15,500	0.0120		15,779,100
16,000	0.0125		15,620,300
500		0.0001	
16,500	0.0130		15,504,000
500		0.0001	

Tensile strength, 28,750 lbs. per sq. inch.

This piece broke first with a load of 27,100 lbs., exhibiting a flaw; upon being re-tested it broke with 30,450 lbs., or 28,750 lbs. per sq. inch

TEST NO. 10.
UNPLANED GUN IRON.
Gauged length, 13".4844.
Area of section, 1.0620 sq. in.

Loads Applied.	Elongation, Inches.	Sets, Inches.	E
500	0.0000		
1,000	0.0C03		21,168,200
1,500	0.0006		22,088,600
2,000	0.0009		21,773,000
2,500	0.0012		21,618,630
3,000	0.0015		20,821,220
3,500	0.0019		20,054,100
4,000	0.0023		19,757,020
4,500	0.0027		18,816,200
5,000	0.0031		18,586,800
5,500	0.0035		18,184,200
6,000	0.0040		17,684,900
6,500	0.0044		17,518,545
7,000	0.0048		17,380,200
7,500	0.0052		17,180,000
500		0.0001	
8,000	0.0057		16,859,700
8,500	0.0061		16,657,000
500		0.0002	

Tensile strength, 28,775 lbs. per sq. inch.

The piece broke first at 23,000 lbs., exhibiting a flaw, and on being re-tested it broke at 30,550 lbs., or 28,775 lbs. per sq. inch.

TEST NO. 11,
PLANED COMMON IRON.
Gauged length, 13".5274.
Area of section, 0.9937 sq. in.

Loads Applied.	Elongation, Inches.	Sets, Inches.	E
500	0.0000		
1,000	0.00C4		17,016,454
1,500	0.0009		15,125,737
500		0.0000	
2,000	0.0014		14,585,532
2,500	0.0021		12,964,917
3,000	0.0028		12,154,610
500		0.0002	
3,500	0.0032		12,762,340
4,000	0.0040		11,911,517
500*		0.0004	
4,500	0.0044		12,375,603
500		0.0015	
5,000	0.0049		12,501,885
500		0.0015	
5,500	0.0056		12,154,610
500		0.0015	
6,000	0.0064		11,698,812
500		0.0018	
6,500	0.0068		12,011,614
500		0.0019	

A load of 10,000 lbs. was left upon this piece over night.

* Tightened the clamps.

TEST NO. 12.

SAME SPECIMEN RE-TESTED.

Loads Applied.	Elongation, Inches.	Sets, Inches.	E
500	0.0000		
1,000	0.0004		18,150,700
1,500	0.0010		14,329,500
2,000	0.0016		13,173,900
2,500	0.0021		12,812,300
500		— 0.0001	
3,000	0.0029		11,941,250
3,500	0.0034		12,039,200
4,900	0.0039		12,216,800
500		0.0000	
4,500	0.0045		12,236,430
500		— 0.0001	
5,000	0.0051		12,070,700
500		0.0000	
5,500	0.0059		11,685,100
500		0.0001	
6,000	0.0064		11,790,600
500		0.0001	
6,500	0.0071		11,585,600
500		0.0000	
7,000	0.0078		11,344,200
500		0.0000	
7,500	0.0085		11,343,800
500		0.0000	
8,000	0.0093		11,037,600
500		0.0001	
8,500	0.0098		11,169,680
500		0.0001	
9,000	0.0106		10,890,430
500		0.0003	
9,500	0.0112		10,939,000
500		0.0003	
10,000	0.0121		10,732,250
500		0.0004	
10,500	0.0129		10,593,800
500		0.0006	
11,000	0.0139		10,320,350
500		0.0006	
11,500	0.0148		10,100,730
500		0.0009	
12,000	0.0157		9,971,330
500		0.0013	

A load of 14,000 lbs. was left upon this piece over night.

TEST NO. 13.

SAME SPECIMEN RE-TESTED A SECOND TIME.

Loads applied.	Elongation, inches.	Sets, inches.	E.
500	0.0000		
1.000	0.0005		13,613,040
1,500	0.0011		12,964,800
2,000	0.0017		12,375,500
2,500	0.0024		11,585,540
3,000	0.0029		11,635,100
3,500	0.0035		11,585,540
4,000	0.0042		11,480,900
4,500	0.0048		11,463,600
5,000	0.0054		11,344,200
5,500	0.0061		11,250,440
6,000	0.0067		11,216,780
6,500	0.0075		10,900,460
7,000	0.0082		10,790,810
7,500	0.0091		10,529,420
8,000	0.0097		10,498,500
500		0.0001	
8,500	0.0105		10,421,500
9,000	0.0112		10,377.650
9,500	0.0119		10,274,000
10,000	0.0128		10,151,000
500		0.0002	
10,500	0.0135		10,121,200
11,000	0.0143		10,030,500
500		0.0002	
11,500	0.0151		9,936,500
12,000	0.0160		9,789,680
500		0.0003	
12,500	0.0170		9,637,510
500		0.0004	
13,000	0.0178		9,586,642
500		0.0004	
13,500	0.0185		9,591,840
500		0.0005	
14,000	0.0192		9,589,220
500		0.0006	
14,500	0.0206		9,270,480
500		0.0010	

Tensile strength, 20,800 lbs. per sq. in.

TEST NO. 14.

PLANED COMMON IRON.

Gauged length, 13."461.
Area of section, 0. 9852 sq. in.

Loads applied.	Elongation, inches.	Sets, inches.	E.
500	0.0000		
1,000	0.0004		10,518,890
1,500	0.0008		18,217,600
2,000	0.0012		17,071,150
2,500	0.0018		15,615,100
3,000	0.0024		14,232,000
3,500	0.0030		13,663,200
500		0.0000	
4,000	0.0037		12,837,900
4,500	0.0043		12,710,000
500		0.0002	
5,000	0.0048		12,742,900
500		0.0002	
5,500	0.0054		12,769,360
500		0.0004	
6,000	0.0060		12,629,900
500		0.0009	
6,500	0.0068		12,145,100
500		0.0010	
7,000	0.0075		11,802,000
500		0.0011	
7,500	0.0083		11,628,300
500		0.0013	
8,000	0.0089		11,481,700
500		0.0014	
8,500	0.0096		11,386,000
500		0.0018	
9,000	0.0102		11,430,200
500		0.0019	
9,500	0.0115		11,028,600
500		0.0021	
10,000	0.0122		10,661,250
500		0.0023	

TEST NO. 15.

SAME SPECIMEN RE-TESTED.

Loads applied	Elongation, inches.	Sets, inches.	E.
500	0.0000		
1,000	0.0004		18,217,400
1,500	0.0008		17,630,000
2,000	0.0013		16,777,700
2,500	0.0018		14,073,400
3,000	0.0024		14,085,46;
3,500	0.0031		13,439,24(
4,000	0.0036		13,101,26(
4,500	0.0042		12,935,59(
5,000	0.0048		12,809,27(
500		−0.0001	
5,500	0.0055		12,421,100
6,000	0.0061		12,319,300
500		0.0000	
6,500	0.0067		12,190,230
7,000	0.0074		12,083,110
7,500	0.0081		11,855,200
8,000	0.0088		11,600,850
500		−0.0001	
8,500	0.0096		11,356,700
9,000	0.0103		11,248,170
9,500	0.0110		11,153,450
500		0.0000	
10,000	0.0119		10,930,510
500		0.0001	
10,500	9.0129		10,612,200
500		0.0003	
11,000	0.0139		10,321,150
500		0.0005	
11,500	0.0148		10,138,000
500		0.0007	
12,000	0.0162		9,714,190

TEST NO. 16.

SAME SPECIMENS RE-TESTED A SECOND TIME.

Loads, Applied.	Elongation, Inches.	Sets, Inches.	E.
500	0.0000		
1,000	0.0005		13,663,220
1,500	0.0012		11,881,090
2,000	0.0017		12,421,100
2,500	0.0023		12,145,100
3,000	0.0028		12,191,340
3,500	0.0035		11,881,090
4,000	0.0043		11,252,060
4,500	0.0050		10,980,600
5,000	0.0056		10,979,870
500		0.0001	
5,500	0.0063		10,930,600
6,000	0.0070		10,812,620
6,500	0.0077		10,646,700
7,000	0.0084		10,572,730
500		0.0002	
7,500	0.0090		10,626,950
8,000	0.0097		10,564,340
8,500	0.0104		10,510,160
9,000	0.0111		10,486,400
500		0.0001	
9,500	0.0118		10,465,440
10,000	0.0125		10,425,750
10,500	0.0132		10,331,350
11,000	0.0140		10,247,410
500		0.0001	
11,500	0.0149		10,086,900
12,000	0.0158		9,976,810
500		0.0002	
12,500	0.0164		10,012,785
500		0.0001	
13,000	0.0174		9,843,800
500		0.0002	
13,500	0.0182		9,759,440
500		0.0002	
14,000	0.0191		9,682,590
500		0.0005	
14,500	0.0202		9,430,500
500		0.0010	

Tensile strength, 20,300 lbs. per square inch.

TEST NO. 17.

PLANED COMMON IRON.

Gauged length, 13".582.
Area of section, 0.996 square inches.

Loads, Applied.	Elongation, Inches.	Sets, Inches.	E.
500	0.0000		
1,000	0.0005		15,151,720
1,500	0.0009		16,054,000
2,000	0.0014		14,6'0,600
2,500	0.0019		14,354,260
3,000	0.0025		13,636,350
· 3,500	0.0032		12,987,200
4,000	0.0037		12,899,440
500		0.0000	
4,500	0.0044		12,114,810
5,000	0.0051		11,973,550
5,500	0.0059		11,605,570
6,000	0.0066		11,363,800
500		0.0006	
6,500	0.0071		11,605,570
7,000	0.0078		11,437,100
500		0.0008	
7,500	0.0084		11,429,200
500		0.0010	
8.000	0.0091		11,238,900
500		0.0014	
8,500	0.0102		10,748,000
500		0.0016	
9,000	0.0109		10,634,000
500		0.0018	
9,500	0.0116		10,580,080
500		0.0021	
10,000	0.0123		10,575,800

2

TEST NO. 18.

THE SAME RE-TESTED.

Loads, Applied.	Elongation, Inches.	Sets, Inches.	E.
500	0.0000		
1,000	0.0005		15,151,720
1,500	0.0009		15,151,720
2,000	0.0014		15,151,720
2,500	0.0019		14,742,215
3,000	0.0024		14,204,740
3,500	0.0030		13,867,700
4,000	0.0036		13,444,480
4,500	0.0042		13,143,660
5,000	0.0049		12,523,360
500		−0.0001	
5,500	0.0056		12,285,200 ·
6,000	0.0063		12,000,170
500		−0.0002	
6,500	0.0070		11,688,500
7,000	0.0077		11,586,600
500		−0.0002	
7,500	0.0084		11,431,840
8,000	0.0092		11,177,500
500		−0.0002	
8,500	o.0099		11,019,430
500		−9.0002	
9,000	0.0106		10,935,000
500		−0.0001	
9,500	0.0114		10,813,120
500		−0.0001	
10,000	0.0121		10,706,130
500		0.0001	
10,500	0.0129		10,611,860
500		0.0002 ·	
11,000	0.0139		10,301,000
500		0.0005	
11,500	0.0150		10,000,000
500		0.0009	
12,000	0.0162		9,710,230

TEST NO. 19.

SAME SPECIMEN RE-TESTED A SECOND TIME.

Loads applied.	Elongation, inches.	Sets, inches.	E.
500	0.0000		
1,000	0.0006		13,396,900
1,500	0.0011		12,987,200
2,000	0.0017		12,396,900
2,500	0.0023		12,121,380
3,000	0.0029		11,755,600
3,500	0.0035		11,688,500
4,000	0.0041		11,640,900
4,500	0.0048		11,483,400
5,000	0.0054		11,363,800
5,500	0.0061		11,288,520
6,000	0.0068		11,029,560
6,500	0.0076		10,337,000
7,000	0.0083		10,679,200
7,500	0.0091		10,587,600
8,000	0.0098		10,424,120
500		0.0000	
8,500	0.0105		10,437,050
9,000	0.0114		10,212,400
500		0.0000	
9,500	0.0121		10,198,700
10,000	0.0129		10,042,400
500		0.0000	
10,500	0.0138		9,917,490
500		0.0000	
11,000	0.0146		9,840,810
500		0.0001	
11,500	0.0153		9,836,200
500		0.0002	
12,000	0.0160		9,832,000
500		0.0001	
12,500	0.0171		9,597,570
500		0.0001	
13,000	0.0178		9,576,230
500		0.0002	
13,500	0.0189		9,379,640
500		0.0004	
14,000	0.0198		9,279,650

Tensile strength, 20,450.

TEST NO. 20.

PLANED GUN IRON.

Gauged length, 18."2774
Area of section, 1.0028 sq. in.

Loads applied.	Elongation, inches.	Sets, inches.	E.
500	0.0000		
1,000	0.0004		18.879,100
1,500	0.0007		18,879,100
2,000	0.0011		18,055,000
2,500	0.0015		17,603,000
3,000	0.0019		17,195,100
3,500	0.0024		16,724,600
4,000	0.0028		16,403,950
4,500	0.0033		16,171,410
5,000	0.0038		15,888,400
500		0.0002	
5,500	0.0041		16,043,900
6,000	0.0046		16,004,800
6,500	0.0050		15,888,400
7,000	'0.0055		15,791,200
500		0.0002	
7,500	0.0059		15,642,550
500		0.0002	
8,000	0.0066		15,150,680
8,500	0.0070		15,240,080
500		0.0007	
9,000	0.0073		15,523,100
9,500	0.0077		15,576,800
500		0.0007	
10,000	0.0081		15,529,800
500		0.0008	
10,500	0.0086		15,395,800
500		0.0009	
11,000	0.0091		15,277,300
500		0.0009	
11,500	0.0097		15,092,600
500		0.0011	
12,000	0.0102		15,075,600
500		0.0013	

Tensile strength, 29,500.

TEST NO. 21.

SAME SPECIMEN RE-TESTED.

Loads applied.	Elongation, inches.	Sets, inches.	E.
500	0.0000		
1,000	0.0004		17,653,780
1,500	0.0008		17,653,780
2,000	0.0012		17.270,000
2,500	0.0016		17,084,320
3,000	0.0020		16,974,800
3,500	0.0024		16,550,400
4,000	0.0029		16,260,000
4,500	0.0033		16,295,800
5,000	0.0037		16,323,700
5,500	0.0041		16,146,740
500		0.0000	
6,000	0.0046		16,004,800
6,500	0.0050		15,888,400
7,000	0.0055		15,792,600
7,500	0.0059		15,708,860
500		0.0001	
8,000	0.0064		15,576,850
8,500	0.0069		15,463,150
9,000	0.0073		15,469,800
500		0.0001	
9,500	0.0078		15,277,660
10,000	0.0083		15,246,440
10,500	0.0087		15,218,770
500		0.0002	
11,000	0.0092		15,111,200
11,500	0.0097		15,014,800
500		0.0002	
12,000	0.0102		14,927,800
12,500	0.0107		14,849,000
500		0.0004	
13,000	0.0113		14,711,480
13,500	0.0117		14,711,480
500		0.0004	
14,000	0.0122		14,711,480
14,500	0.0128		14,481,600
500		0.0006	

TEST NO. 22.

SAME SPECIMEN RE-TESTED A SECOND TIME.

Loads Applied.	Elongation, Inches.	Sets, Inches.	E.
500	0.0000		
1,000	0.0004		17,653,900
1,500	0.0008		17,653,900
2,000	0.0012		17,653,900
2,500	0.0016		17,084,300
3,000	0.0020		16,759,900
3,500	0.0024		16,550,410
4,000	0.0028		16,404,000
4,500	0.0033		16,295,800
5,000	0 0037		15,995 000
5,500	0.0042		15,856,700
6,000	0.0046		15,827,200
6,500	0.0051		15,653,600
7,000	0.0056		15,506,700
500		0.0005	
7,500	0.0061		15,319,400
8,000	0.0065		15,395,700
8,500	0.0068		15,576,890
9,000	0.0072		15,740,300
9,500	0.0076		15,783,170
10,000	0.0080		15,722,900
500		0.0001	
10,500	0.0088		15,131,800
11,000	0.0093		15,029,600
11,500	0.0098		14,937.800
12,000	0.0103		14,855,000
560		0.0001	
12,500	0.0109		14,644,000
13,000	0.0113		14.646,720
13,500	0.0118		14,586,800
500		0.0002	
14,000	0.0123		14.532,090
14,500	0.0128		14,152,000
500		0.0002	
15,000	0.0134		14,327,220
500		0.0002	
15,500	0.0140		14,160,780
500		0.0003	
16,000	0.0146		14,056,500
500		0.0004	
16,500	0.0152		13,937,200
500		0.0006	
17,000	0.0158		13,870,800
500		0.0009	

Tensile strength, 29,500.

TEST NO. 23.

PLANED GUN IRON.

Gauged length, 13.508 inches.
Area of section, 0.9930 square inches.

Loads Applied.	Elongation, Inches.	Sets, Inches.	E.
500	0.0000		
1,000	0.0003		20,927,000
1,500	0.0007		20,927,000
2,000	0.0010		20,404,350
2,500	0.0014		19,786,500
3,000	0.0018		19,433,200
3,500	0.0021		19,204,560
4,000	0.0025		18,856,000
4,500	0.0030		18,602,700
5,000	0.0034		18,273,000
500		0.0000	
5,500	0.0039		17,440,000
6,000	0.0044		17,199,500
6,500	0.0048		17,004.000
7,000	0.0053		16,684,000
500		0.0000	
7,500	0.0057		16,632,800
8,000	0.0062		16,455,550
8,500	0.0067		16,242,650
500		0.0002	
9,000	0.0071		16,228,400
9,500	0.0077		16,003,800
500		0.0004	
10,000	0.0081		16,053,500
500		0.0005	
10,500	0.0087		15,726,270
500		0.0007	
11,000	0.0090		15,870,430
500		0.0009	
11,500	0.0095		15,834,450
500		0.0012	
12,000	0.0099		15,881,940

TEST No. 24.

SAME SPECIMEN RE-TESTED.

Loads Applied.	Elongation, Inches.	Sets, Inches.	E.
500	0.0000		
1,000	0.0004		19,433,200
1,500	0.0007		18,763,070
2,000	0.0011		18,549,900
2,500	0.0016		17,552,550
500		0.0001	
3,000	0.0020		17,219,270
3,500	0.0024		17,004,030
4,000	0.0029		16,705,700
4,500	0.0033		16,488,760
5,000	0.0038		16,328,860
5,500	0.0042		16,194,300
6,000	0.0047		16,089,900
6,500	0.0051		16,003,800
7,000	0.0056		15,931,740
7,500	0.0060		15,870,430
8,000	0.0065		15,817,700
500		0.0001	
8,500	0.0070		15,858,400
9,000	0.0074		15,625,330
9,500	0.0079		15,596,053
10,000	0.0083		15,569,950
500		0.0002	
10,500	0.0088		15,546,540
11,000	0.0092		15,525,420
11,500	0.0097		15,426,340
500		0.0004	
12,000	0.0101		15,527,250
500		0.0004	
12,500	0.0106		15,472,900
500		0.0005	
13,000	0.0110		15,459,200
500		0.0005	
13,500	0.0116		15,311,000
500		0.0006	
14,000	0.0121		15,240,130

TEST NO. 25

SAME SPECIMEN RE-TESTED A SECOND TIME.

Loads Applied.	Elongation, Inches.	Sets, Inches.	E
500	0.0000		
1,000	0.0004		18,137,700
1,500	0.0008		18,137,700
2,000	0.0012		17,743,340
2,500	0.0016		17,552,550
3,000	0.0020		17,440,030
3,500	0.0023		17,740,340
4,000	0.0027		17,633,800
4,500	0.0031		17,552,550
5,000	0.0035		17,489,900
5,500	0.0040		17,219,280
6,000	0.0044		17,004,030
6,500	0.0049		16,828,740
7,000	0.0053		16,682,200
7,500	0.0058		16,556,560
8,000	0.0062		16,455,500
8,500	0.0067		16,364,800
9,000	0.0071		16,285,530
9,500	0.0076		16,215,260
10,000	0.0080		16,153,800
500		0.0000	
10,500	0.0085		16,051,000
11,000	0.0089		15,959,100
11,500	0.0094		15,876,440
12,000	0.0099		15,801,730
500		0.0000	
12,500	0.0105		15,620,700
13,000	0.0112		15,250,250
500		0.0000	
13,500	0.0118		14,990,060
14,000	0.0123		14,930,375
500		0.0000	
14,500	0.0128		14,849,530
500		0.0001	
15,000	0.0133		14,886,550
500		0.0001	
15,500	0.0138		14,839,900
500		0.0002	
16,000	0.0143		14,796,500
500		0.0003	
16,500	0.0149		14,583,025
500		0.0005	
17,000	0.0156		14,388,000

Tensile strength, 31,000.

TEST NO. 26.

PLANED GUN IRON.

Gauged length, 13″.5274.

Area of section, 0.99 sq. inch.

Loads Applied.	Elongation, Inches.	Sets, Inches.	E
500	0.0000		
1,000	0.0003		21,021,600
1,500	0.0007		21,021,600
2,000	0.0010	と ৷,	21,021,600
2,500	0.0013		20,625,000
3,000	0.0017		20,694,200
3,500	0.0021		19,520,050
4,000	0.0025		19,129,700
4,500	0.0029		18,847,000
5,000	0.0034		18,218,320
500		0.0000	
5,500	0.0039		17,681,020
6,000	0.0044		17,276,400
6,500	0.0048		17,080,000
7,000	0.0053		16,917,400
7,500	0.0057		16,780,400
500		0.0001	
8,000	0.0062		16,529,100
8,500	0.0067		16,315,300
500		0.0003	
9,000	0.0071		16,301,000
9,500	0.0077		16,075,320
500		0.0004	
10,000	0.0082		15,927,400
500		0.0006	
10,500	0.0086		15,888,420
500		0.0007	
11,000	0.0093		15,510,530
500		0.0009	
11,500	0.0098		15,415,850
500		0.0010	
12,000	0.0101		15,558,060

TEST NO. 27.

SAME SPECIMEN RE-TESTED.

Loads Applied.	Elongation, Inches.	Sets, Inches.	E
500	0.0000		
1,000	0.0004	.	19,520,060
1,500	0.0007		19,520,060
2,000	0.0011		19,066,100
2,500	0.0015		18,846,950
3,000	0.0019		18,218,720
3,500	0.0023		17,882,670
4,000	0.0028		17,390,600
4,500	0.0082		17,351,170
5,000	0.0086		17,320,600
5,500	0.0040		17,180,710
500		0.0001	
6,000	0.0044	'	17,177,500
6,500	0.0049		16,605,935
7,000	0.0054		16,447,460
7,500	0.0059		16,211,940
8,000	0.0064		16,059,445
500		0.0002	
8,500	0.0070		15,728,400
9,000	0.0074		15,695,800
9,500	0.0078		15,766,200
500		0.0002	
10,000	0.0083		15.592,600
10,500	0.0088		15,616,050
11,000	0.0093		15,510,540
500		0.0002	
11,500	0.0097		15,455,471
12,000	0.0102		15,367,850
500		0.0004	
12,500	0.0106		15,468,700
500		0.0004	
13,000	0 0111		15,387,440
500		0.0004	
13,500	0.0116		15,280,340
500		0.0004	
14,000	0.0122		15,182,270
500		0.0005	

Tensile strength, 31,000.

TEST NO. 28.

SAME SPECIMEN RE-TESTED A SECOND TIME.

Loads Applied.	Elongation, Inches.	Sets, Inches.	E.
500	0.0000		
1,000	0.0004		18,176,400
1,500	0.0008		18,176.400
2,000	0.0012		17,781,700
2,500	0.0016		17,714,500
3,000	0.0020		16,830,400
3,500	0.0025		16,524.400
4,000	0.0029		16,312,530
4,500	0.0034		16,120,000
5,000	0.0038		16,038,370
5,500	0.0043		15,907,900
6,000	0.0047		15,868,650
6,500	0.0052		15,806,000
7,000	0.0056		15,753,250
7,500	0.0061		15,708,360
8,000	0.0066		15.609,500
8,500	0.0070		15 524,700
9,000	0.0075		15,450,200
9,500	0.0080		15.336,700
10,000	0.0086		14,999,200
500		0.0002	
10,500	0.0091		14,980.900
11,000	0.0095		15,067.620
11,500	0.0100		15.071,200
12,000	0 0104		15,074,530
12,500	0.0109		15,008.380
13,000	0.0114		15,013.900
13,500	0.0119		14.955,600
14,000	0.0124		14,914,100
14,500	0.0129		14,852,750
500		0.0003	
15,000	0.0134		14,724,250
15,500	0.0140		14,580,330
16,000	0.0146		14,448,250
500			
16,500	0.0152		14,346,820
17,000	0.0158		14,259,150
500		0.0005	

Tensile strength, 31,000 lbs. per square inch.

From these tests Mr. Cochran obtains the following as average values for the specimens tested, viz.:

For tensile strength :

Unplaned common	22,066
Planed common	20,520
Unplaned gun	28,175
Planed gun	30,500

For limit of elasticity :

Unplaned common	6,500
Planed common	5,833
Unplaned Gun	11,000
Planed Gun	8,500

For modulus of elasticity at assumed elastic limit :

Unplaned common	13,194,233
Planed common	11,943,953
Unplaned gun	16,180,300
Planed gun	15,932,880

Colonel Rosset of the Turin arsenal gave for gun iron, as average limit of elasticity 9,800, and as average modulus of elasticity 16,263,300.

He attributes the apparent anomaly in the case of gun iron, whose average tensile strength is less in the unplaned than in the planed, to the presence of surface flaws in the unplaned gun.

He draws from his tests the following conclusions, viz.:

1°. Planed pieces stretch more than unplaned.

2°. The moduli of planed are higher than those of unplaned pieces.

3°. Common iron stretches from $\frac{1}{4}$ to $\frac{1}{2}$ more than gun iron.

4°. The elastic limit for unplaned is higher than that for planed.

5°. The effect of re-testing is to lower the modulus of elasticity, to raise the elastic limit, to make the stretch more nearly equal on the two sides, and probably to lower the tensile strength.

SUMMARY OF THE EXPERIMENTS OF MESSRS. BURGESS AND VIELÉ.

The object of this investigation was to determine the transverse strength of cast iron in the form of window lintels, and also the deflections under moderate loads, and from the latter to deduce the modulus of elasticity of the cast iron, and to compare it with the modulus of elasticity of the same iron, as determined from tensile experiments ; also the tensile strength and limit of elasticity of specimens taken from different parts of the lintel were determined.

The iron used was of two qualities, marked P and S respectively ; that marked P was composed of what was called at the foundry of L. M. Ham & Co., where the casting was done, No. 1 and No. 2 pig.

No. 1 pig was prepared by mixing the following ores :'

```
Neshannock from Pennsylvania............................25%
Franklin from New York..................................37.5%
Crozen from Virginia....................................37.5%
```

No. 2 pig was made by mixing Franklin and Crozen in equal parts.

The chemical composition of *P* is as follows:

```
Graphite..................... ........................... 3.00
Combined carbon............................... ........... 0.56
Sulphur................... ............................... 0.53
Silicon.................................................. 1.34
Phosphorus........... ..................................... 1.13
Manganese.................... ............................ 0.33
Iron by difference....................................... 93.11
```

The iron marked *S* was made of old scrap. Its chemical composition was as follows :

```
Graphite................................................. 2.39
Combined carbon.......................................... 0.85
Sulphur.................................................. 0.07
Silicon.................................................. 1.49
Phosphorus............................................... 1.12
Manganese................................................ 0.40
Iron by difference.......................................93.68
```

The specimens for tension were 24 inches long, and about one inch square in section.

The transverse tests were made on window lintels of the following dimensions :

```
                                                          Inches.
Length....................................................  54
Breadth of flange........................................   8
Height of web at the centre of lintel above flange...........   4
Height of web at edge of lintel above flange................  2.5
Thickness of web and flange...............................  0.75
```

The tensile specimens were cast at the same time, and from the same run as the lintels.

Besides this, one of each kind of window lintels was cut up into tensile specimens, and the specimens were so marked as to show from what part of the lintel they were cut.

The tables of tests will now be given, and the following explanation of the symbolism employed.

P and *S* are used, as already stated, to denote the quality of the iron.

A and *B* are used to denote respectively that the specimen was unplaned or planed.

1, 2, 3, etc., denote the number of the test made on that particular kind and condition.

I., II., III., denote that the piece has been taken from a lintel, and also from what part, as will easily be seen by the accompanying sketch (Fig. 31.)

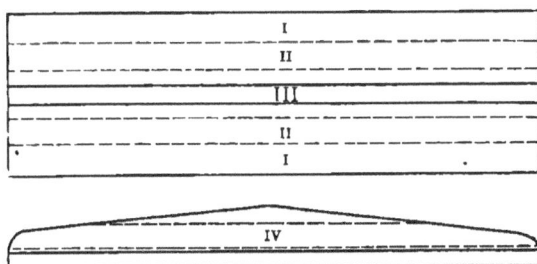

Fig. 31.

Thus *P. B.* 3 would signify that the specimen was of quality *P*, had been planed, and was the third test of this class. ·

On the other hand, *P. B.* 3 II., would signify in addition that it had been taken from a lintel, and was a piece of one of the strips marked II. in the sketch.

Loads applied	P.A.1 Area = 1.0568 sq. in.			S.A.1 Area = 0.9688 sq. in.			S.A.2 Area = 1.0302 sq. in.			P.A.2 Area = 1.0353 sq. in.			S.A.3 Area = 1.02 sq. in.			P.A.3 Area = 1.027 sq. in.		
	Loads per sq. in.	Elongations, inches.	E.	Loads per sq. in.	Elongations, inches.	E.	Loads per sq. in.	Elongations, inches.	E.	Loads per sq. in.	Elongations, inches.	E.	Loads per sq. in.	Elongations, inches.	E.	Loads per sq. in.	Elongations, inches.	E.
500	473	0.0000		516	0.0000		485	0.0000		484	0.0000		490	0.0000		457	0.0000	
1,000										969	0.0002	24,224,645				974	0.0003	16,225,496
1,500										1,453	0.0004	24,224,645				1,461	0.0006	12,982,708
2,000										1,937	0.0007	21,538,186				1,947	0.0012	13,291,913
2,500	2,365	0.0014	14,002,664	2,582	0.0011	18,777,052				2,422	0.0010	20,187,357	2,451	0.0011	17,825,311	2,434	0.0015	13,446,471
3,000							2,912	0.0010	24,265,138	2,906	0.0014	18,918,485				2,921	0.0018	14,002,876
3,500										3,875	0.0016	18,995,337				3,404	0.0022	13,987,430
4,000										4,359	0.0019	18,568,841				3,895	0.0026	13,534,180
4,500	4,257	0.0027	14,271,942	4,647	0.0022	18,777,052				4,744	0.0024	17,611,059	4,412	0.0022	18,249,729	4,382	0.0030	13,581,708
5,000										5,228	0.0028	17,000,099				4,869	0.0034	13,425,004
5,500							5,339	0.0023	21,121,593	5,812	0.0032	16,511,331				5,355	0.0037	13,705,353
6,000										6,297	0.0036	16,111,430				5,842	0.0040	13,442,958
6,500	6,150	0.0044	13,115,313	6,713	0.0034	18,504,992				6,781	0.0040	15,578,190	6,373	0.0033	17,849,914	6,329	0.0043	13,336,992
7,000										7,265	0.0045	14,985,353				6,816	0.0048	13,247,329
7,500										7,750	0.0049	14,704,101				7,303	0.0053	12,906,500
8,000							7,766	0.0039	19,299,067	8,234	0.0054	14,458,006				7,790	0.0057	12,941,587
8,500	8,042	0.0062	12,636,999	8,778	0.0047	17,850,759				8,718	0.0058	14,240,805	8,333	0.0048	17,309,104	8,276	0.0062	12,741,304
9,000										9,203	0.0063	13,988,034				8,763	0.0067	12,628,527
9,500										9,687	0.0068	13,612,055				9,260	0.0069	12,377,451
10,000										10,171	0.0072	13,383,348				9,335	0.0073	12,366,604
10,500	9,934	0.0040	11,889,057	10,843	0.0061	17,129,739	10,192	0.0054	18,519,273	10,656	0.0077	13,215,465	10,294	0.0062	16,461,582	10,234	0.0083	12,153,986
11,000										11,140	0.0081	13,062,181				10,711	0.0087	12,154,814
11,500										11,624	0.0093	12,921,669				11,198	0.0097	11,971,152
12,000										12,109	0.0098					11,684	0.0103	11,970,857
12,500				12,909	0.0076	16,569,581							12,255	0.0076	16,052,252	12,171	0.0107	11,818,791
13,000							12,619	0.0073	17,369,853							12,658	0.0111	11,689,466
13,500																13,145	0.0118	11,580,466
14,000													14,216	0.0091	15,626,487	13,682	0.0124	11,393,945
14,500																		
15,000																		
15,500							15,046	0.0093	16,547,325									
16,000																		
16,500													16,177	0.109	15,084,883			

	P.A.1	S.A.1	S.A.2	P.A.2	S.A.3	P.A.3
Breaking load	25,110	23,440	26,020	22,115	25,200	19,450
Per sq. in.	23,757	24,204	25,258	21,423	24,706	18,308

Loads Applied	P. A. 4 Area = 1.0514			P. B. 1 Area = 0.828			S. B. 1 Area = 0.946			P. B. 3 Area = 0.9025			P. B. 4 II. Area = 0.7398			S. B. 2 Area = 0.887		
	Loads per sq. in.	Elong. inches	E.	Loads per sq. in.	Elong. inches	E.	Loads per sq. in.	Elong. inches	E.	Loads per sq. in.	Elong. inches	E.	Loads per sq. in.	Elong. inches	E.	Loads per sq. in.	Elong. inches	E.
500	476	0.0000		604	0.0000		591	0.0000		554	0.0000		658	0.0000		597	0.0000	
1,500	1,427	0.0004	23,777,819	1,812	0.0007	17,253,675	1,773	0.0005	23,640,661	1,662	0.0006	20,146,060	1,974	0.0007	18,901,511	1,792	0.0007	17,067,759
2,500	2,378	0.0013	17,172,809	3,020	0.0014	17,253,675	2,955	0.0011	21,670,604	2,770	0.0013	17,459,918	3,290	0.0018	15,383,177	2,987	0.0014	17,067,759
3,500	3,329	0.0017	19,374,510	4,226	0.0023	16,238,618	4,137	0.0018	20,508,770	3,878	0.0022	15,742,772	4,606	0.0031	13,630,160	4,181	0.0021	17,067,759
4,500	4,280	0.0025	17,701,965	5,436	0.0034	14,023,903	5,319	0.0026	18,558,150	4,986	0.0030	15,501,873	5,922	0.0046	12,416,202	5,376	0.0028	17,067,759
5,500	5,231	0.0033	16,589,791	6,614	0.0046	13,951,927	6,501	0.0033	18,463,757	6,094	0.0041	14,415,624	7,238	0.0062	11,631,163	6,571	0.0036	16,041,065
6,500	6,182	0.0040	16,046,883	7,832	0.0050	13,639,488	7,683	0.0041	17,848,999	7,202	0.0051	13,771,804	8,554	0.0079	10,946,098	7,766	0.0044	16,356,602
7,500	7,133	0.0048	15,852,856	9,060	0.0070	12,993,364	8,865	0.0049	17,409,886	8,310	0.0062	13,311,931	9,870	0.0102	10,218,014	8,960	0.0053	16,027,916
8,500	8,084	0.0063	14,313,909	10,288	0.0085	12,314,381	10,047	0.0059	16,711,192	9,418	0.0073	12,853,324	11,186	0.0127	9,598,935	10,155	0.0062	15,596,457
9,500	9,035	0.0070	14,132,495	11,476	0.0101	11,759,401	11,229	0.0069	16,326,888	10,526	0.0086	12,371,825	12,502	0.0155	9,045,518	11,350	0.0071	15,425,273
10,500	9,987	0.0080	13,570,359	12,684	0.0119	11,273,592	12,411	0.0079	15,795,232	11,634	0.0095	11,767,354	13,818	0.0192	8,501,551	12,544	0.0081	15,020,599
11,500	10,938	0.0091	14,190,164	13,892	0.0138	10,826,561	13,593	0.0088	15,490,434	12,742	0.0115	11,573,510				13,739	0.0092	14,689,502
12,500	11,889	0.0101	12,939,411	15,100	0.0160	10,361,689	14,775	0.0099	15,187,686	13,850	0.0133	11,138,586				14,934	0.0102	14,460,997
13,500	12,840	0.0114	12,506,800	16,308	0.0184	9,970,395	15,957	0.0111	14,700,653	14,958	0.0152	10,728,614				16,128	0.0112	14,223,684
14,500	13,791	0.0124	12,062,043	17,516	0.0212	9,566,257	17,159	0.0123	14,381,789	16,066	0.0171	10,378,638				17,323	0.0123	13,983,702
15,500	14,742	0.0142	11,746,257	18,724	0.0246	9,165,316	18,341			17,174	0.0193	10,015,229				18,518	0.0136	13,688,653
16,500	15,693	0.0159	11,351,799				19,503	0.0148	13,770,617									
Breaking load	22,510			18,010			25,020			22,750			14,450			19,420		
Per sq. in.	21,409			21,756			29,574			25,207			19,016			23,201		

3

Loads applied.	S. B. 3 II. Area = 0.8432 sq. in.			S. B. 4 IV. Area = 1.3478 sq. in.			P. B. 5 I. Area = 0.8879 sq. in.			P. B. 6 I. Area = 0.9664 sq. in.			S. B. 4 II. Area = 0.837 sq. in.			P. B. 7 II. Area = 0.7442 sq. in.		
	Loads per sq. in.	Elongations, inches.	E.	Loads per sq. in.	Elongations, inches.	E.	Loads per sq. in.	Elongations, inches.	E.	Loads per sq. in.	Elongations, inches.	E.	Loads per sq. in.	Elongations, inches.	E.	Loads per sq. in.	Elongations, inches.	E.
500	593	0.0000		371	0.0000		570	0.0000		517	0.0000		597	0.0000		672	0.0000	
1,000																		
1,500	1,779	0.0006	21,562,876	1,113	0.0003	24,731,657	1,709	0.0008	15,169,480	1,552	0.0006	18,814,000	1,792	0.0005	26,549,859	2,016	0.0011	12,737,378
2,000																		
2,500	2,965	0.0018	19,252,569	1,855	0.0008	19,785,320	2,848	0.0018	13,290,801	2,587	0.0014	15,874,302	2,987	0.0011	23,231,120	3,359	0.0022	12,506,529
3,000																		
3,500	4,151	0.0030	18,488,406	2,597	0.0013	18,686,143	3,987	0.0030	12,025,010	3,621	0.0023	14,218,634	4,181	0.0019	21,278,919	4,703	0.0034	11,920,952
4,000																		
4,500	5,337	0.0029	16,968,792	3,339	0.0019	17,106,005	5,126	0.0042	11,297,183	4,656	0.0034	13,123,959	5,376	0.0028	19,473,140	6,047	0.0050	11,040,284
5,010																		
5,500	6,523	0.0038	16,376,724	4,081	0.0023	16,968,416	6,266	0.0056	10,725,467	5,691	0.0045	12,298,765	6,571	0.0036	18,889,673	7,390	0.0066	10,511,883
6,000																		
6,500	7,709	0.0046	16,118,016	4,823	0.0029	16,400,399	7,405	0.0072	10,088,608	6,726	0.0058	11,575,396	7,766	0.0046	17,490,768	8,734	0.0086	9,879,673
7,000																		
7,500	9,995	0.0056	15,509,608	5,565	0.0034	16,177,290	8,544	0.0091	9,508,927	7,760	0.0072	11,016,932	8,960	0.0055	16,728,693	10,078	0.0109	9,321,449
8,000																		
8,500	11,181	0.0065	15,131,431	6,307	0.0039	15,841,378	9,683	0.0112	8,994,039	8,795	0.0086	10,531,857	10,155	0.0065	16,131,035	11,421	0.0136	8,778,363
9,000																		
9,500	12,367	0.0075	14,707,892	7,049	0.0045	15,550,115	10,822	0.0134	8,570,061	9,830	0.0102	10,080,240	11,350	0.0076	15,602,976	12,765	0.0167	8,276,966
10,000																		
10,500	13,553	0.0088	11,239,870	7,790	0.0052	15,082,032	11,962	0.0161	8,134,985	10,564	0.0121	9,631,550	12,514	0.0087	15,128,808	14,109	0.0210	7,765,440
11,000																		
11,500	14,739	0.0100	13,807,851	8,532	0.0059	14,674,509				11,899	0.0141	9,214,832	13,139	0.0099	14,658,571	15,453	0.0256	7,323,194
12,000																		
12,500	15,925	0.0113	13,417,566	9,274	0.0064	14,575,719				12,934	0.0164	8,821,845	13,934	0.0113	14,148,180	16,796	0.0320	6,886,975
13,000																		
13,500				10,016	0.0071	14,332,631				13,968	0.0191	8,443,610	16,139	0.0127	13,716,305			
14,000																		
14,500				10,758	0.0078	14,065,963				15,003	0.0222	8,078,921	17,333	0.0141	13,325,112			
15,000																		
15,500				11,500	0.0085	13,834,918				16,038	0.0262	7,712,788	18,528	0.0159	12,905,298			
16,000																		
16,500				12,242	0.0092	13,588,527												
Breaking load	20,830			34,760			17,250			20,020			21,630			14,420		
Per sq. in.	24,704			25,790			19,651			20,715			29,414			19,376		

Designation of Specimen.	Area sq. in.	Breaking weight.	Breaking wt. per sq. in.	Remarks.
P. B. 2. III.	0.96	12,240	12,754	Broke at a flaw at 10,170 lbs. re-tested and broke at a flaw at 12,240 lbs.
P. B. 8. IV.	1.2276	24,080	19,616	Broke at a slight flaw.
P. B. 9. I.	0.9513	20,050	21,706	
S. B. 5. II.	0.8001	18,890	23,610	
P. B. 10. I.	0.9333	20,050	21,483	
P. B. 11. II.	0.741	16,410	22,146	
S. B. 6. I.	0 8512	24,700	29,124	
P. B. 12. II.	0.725	14,900	20,552	
S. B. 7. I.	0.8385	23,590	28,372	
S. B. 8. I.	0.8645	21,980	25,425	
P. B. 13. III.	0.8624	13,920	16,141	
S. B. 9. III.	1.1063	30,550	27,523	
S. B. 10. III.	1.3275	24,340	18,301	

The following is a summary of the breaking weights of the specimens not cut from the lintels.

P. A. 1..............23,757
P. A. 2..............21,423
P. A. 3..............18,398
P. A. 4..............21,409
————
4)84,987
————
21,247

S. A. 1..............24,204
S. A. 2..............25,258
S. A. 3..............24,706
————
3)74,168
————
24,723

P. B. 1. 21,756
P. B. 3. 25,207
————
2)46,963
————
23,482

S. B. 1..............29,574
S. B. 2..............23,201
————
2)52,775
————
26,388

The conclusions which Messrs. Burgess and Vielé draw from these tests are the following, viz. :

1°. The tensile strength of the iron marked S was higher than that of the iron marked P.

2°. The elongations for a certain load were greater for equal areas with the grade P than with the grade S.

3°. Hence S was a stronger, but, at the same time, a more brittle iron.

4°. With the same grade of iron, the elongations were greater in planed than in unplaned specimens.

5°. The unplaned specimens in these tests had a less tensile strength per square inch than the planed. They attribute this fact to some slight irregularities in the castings, which were removed by planing.

6°. In regard to the tensile specimens cut from the lintels, it will be seen that specimens marked I. and II. broke at higher loads than those marked IV., and that the weakest of all were those marked III.

TESTS OF THE TRANSVERSE STRENGTH OF WINDOW LINTELS.

All the window lintels tested were of the form shown in the cut (Fig. 31), and all were supported at the ends and loaded in the middle, the span in every case being 52″. From the cut it will be seen that the web varied in height, being 4 inches high above the flange in the centre, and decreasing to 2.5 inches at the ends over the supports. Inasmuch as the section, and hence the moment of inertia of the section varied, it became necessary to deduce a special approximate formula suitable to determine the modulus of elasticity from the observed deflections.

In order to deduce this special formula, the moments of inertia were first determined at the following five sections, viz.:

Distance of section from support, inches.	Moment of inertia of section about neutral axis.
26	15.5625
19¼	12.2072
13	9.3600
6¼	6.9773
0	5.0300

These five values satisfy very nearly the equation:

$$I = \frac{1.8725}{338} x^2 + \frac{6.7875}{26} x + 5.03$$

Hence this was used for I in the general deflection equation:

$$\frac{d^2 v}{dx^2} = \frac{M}{E\,I}$$

and hence was deduced:

$$E = \frac{321.695 \; W}{v}$$

where W = load applied, and v = resulting deflection.

A perusal of the results will show that the P's which in tension bore the least were in every case the ones which in the form of lintels stood the most. On the whole, the tensile and the compressive moduli of rupture compare very well with the tensile and the compressive strength of the iron respectively.

The results of the separate tests are given in the following tables:

Loads applied.	S. 1. Span 52". Wt. of lintel 119 lbs.		S. 2. Span 52". Wt. of lintel, 116 lbs.		P. 1. Span 52". Wt. of lintel, 119 lbs.		S. 3. Span 52". Wt. of lintel, 117 lbs.	
	Deflect. Inches.	E.	Deflect. Inches.	E.	Deflect. Inches.	E.	Deflect. Inches.	E.
500	0.0000		0.0000					
1,500			0.0147	21,958,707				
2,500	0.0271	23,785,209	0.0296	22,551,118	0.0331	19,467,170	0.0291	22,109,622
3,500			0.0441	21,932,252				
4,500	0.0557	23,121,053	0.0600	21,522,288	0.0693	18,607,930	0.0598	21,533,476
5,500			0.0759	21,264,309				
6,500	0.0853	22,659,404	0.0923	20,979,580	0.1084	17,890,286	0.0907	21,296,211
7,500			0.1072	21,077,206				
8,500	0.1055	22,320,629	0.1244	20,773,681	0.1484	17,443,935	0.1216	21,177,579
9,500			0.1412	20,598,107				
10,500	0.1363	22,034,361	0.1587	20,377,321	0.1937	16,795,722	0.1557	20,715,611
11,500			0.1760	20,210,429				
12,500	0.1684	21,707,729	0.1938	20,036,531	0.2400	16,309,955	0.1899	20,398,439
13,500			0.2122	19,840,139				
14,500	0.2026	21,290,212	0.2331	19,519,799	0.2927	15,724,039	0.2270	19,961,812
15,500								
16,500							0.2659	19,534,035
	Breaking load...... 26,750 Tensile modulus of rupture....... 26,198 Compr. modulus of rupture....... 80,164		Breaking load........ 19,850 Tensile modulus of rupture........ 19,433 Compr. modulus of rupture........ 59,490		Breaking load........ 27,220 Tensile modulus of rupture....... 26,648 Compr. modulus of rupture........ 81,578		Breaking load...... 28,670 Tensile modulus of rupture....... 28,068 Compr. modulus of rupture....... 85,924	

Loads applied.	S. 4. Span 52". Weight of lintel, 118 lbs.		P. 2. Span 52". Weight of lintel, 120 lbs.		P. 3. Span 52". Weight of lintel, 119 lbs.	
	Deflect. Inches.	E.	Deflect. Inches.	E.	Deflect. Inches.	E.
500	0.0000		0.0000		0.0000	
1,500			0.0149	21,590,272		
2,500	0.0287	22,456,900	0.0307	20,975,359	0.0347	18,541,496
3,500			0.0477	20,291,319		
4,500	0.0588	21,808,268	0.0659	19,637,378	0.0724	17,815,104
5,500			0.0846	19,855.172		
6,500	0.0899	21,712,057	0.1022	19,187,035	0.1150	16,952,791
7,500			0.1204	18,953,963		
8,500	0.1210	21,447,685	0.1403	18,610,508	0.1572	16,521,625
9,500			0.1601	18,343,378		
10,500	0.1544	21,016,559	0.1807	18,070,663	0.2024	16,067,318
11,500			0.2016	17,827,158		
12,500	0.1864	20,844,894	0.2251	17,484,756	0.2522	15,540,516
13,500			0.2492	17,166,571		
14,500	0.2202	20,602,958	0.2730	16,905,850	0.3092	14,932,924
15,500			0.2989	16,605,250		
16,500	0.2637	19,876,418			0.3756	14,277,531
	Breaking load. 25,120 Tensile modulus of rupture .... 24,592 Compr. modulus of rupture....... 75,285		Breaking load. 30,520 Tensile modulus of rupture........ 29,879 Compr. modulus of rupture....... 91,467		Break'g load. 27,200 Tensile modulus of rupture....... 26,659 Compr. modulus of rupture....... 81,608	

SUMMARY OF THE EXPERIMENTS OF MESSRS. EASTMAN AND GERRISH.

The object of this thesis was to determine the constants suitable to use in the formulæ for determining the strength of the arms of cast iron pulleys; and also, incidentally, to determine the holding power of keys and set screws.

Some old pulleys which had been in use at the shops were employed for these tests. They were all about fifteen inches in diameter, and were bored for a shaft $1\frac{3}{16}$ inches in diameter.

Inasmuch as this size of shaft would not bear the strain neces-

Fig: 32

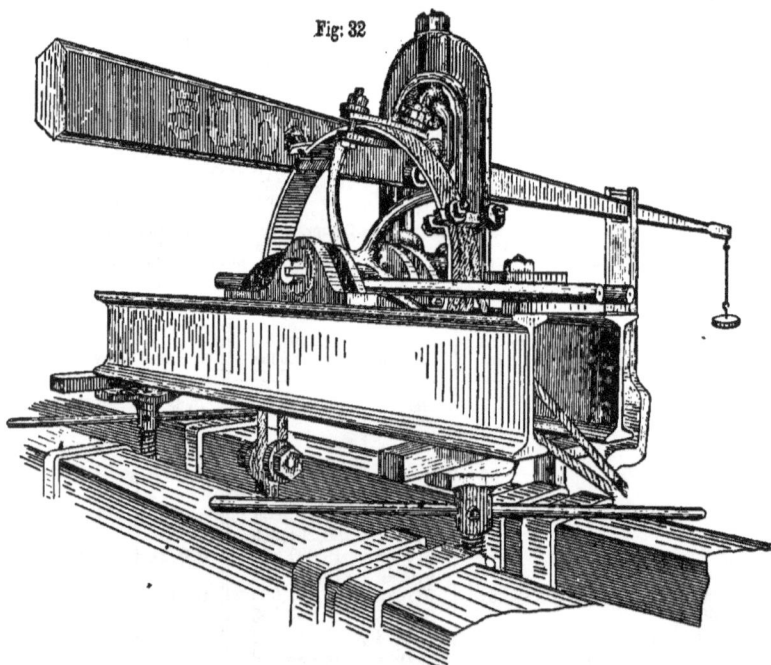

sary to break the arms, the hubs were bored out to a diameter of $1\frac{11}{16}$ inches diameter, and key-seated for a key one-half an inch square.

In order to strengthen the hubs sufficiently, two wrought iron rings were shrunk on them, so as to make it a test of the arms and not of the hub.

The machine used for applying the stress is shown in the cut (Fig. 32).

The pulley under test is keyed to a shaft which, in its turn, is keyed to a pair of castings supported by two wrought iron I beams, resting upon a pair of jackscrews, by means of which the strain is applied. A wire rope is wound around the rim of the pulley, and leaves it in a tangential direction vertically. This rope is connected with the weighing lever of the machine, and weighs the load applied.

The idea of the arrangement was to get a pull upon the rim of the pulley as nearly as possible like the belt pull, to which it is subjected in practice, and, at the same time to have some means of weighing this pull. In practice there are two pulls upon the rim, that of the tight side, and that of the loose side of the belt, the sum of the two tending to produce a bending of the shaft and a compression of the rim and arms of the pulley, while the difference of the two causes a rotation of the pulley and a bending moment in all the arms. It will be seen in the arrangement used that while there is no tight side and loose side of a belt, yet there is a compression of both rim and arms, which must be very similar to that caused by a belt, and a bending moment in the arms such as occurs in practice.

In all the experiments one arm gave way first, and then the unsupported part of the rim broke.

The breaking load of the separate pulleys was, of course, determined, and then it was sought to compute from this the modulus of rupture of the cast iron, if so it can be called. '

The method commonly given for computing the strength of pulley arms is to consider them in one of two ways, viz., either as beams fixed in direction at one end and loaded at the other, or else to consider them as fixed in direction at both ends, thus making of each arm a pair of cantilevers, half as long as the arm, fixed at one end and loaded at the other.

If we let

$I =$ moment of inertia of section,

$n =$ number of arms,

$y =$ half depth of each arm $=$ distance from neutral axis to outside fibre,

$x =$ length of each arm in a radial direction,

$P =$ breaking load determined by experiment:

Then we should have, for the outside fibre stress at fracture,

$$f = \frac{Pxy}{nI} \tag{1}$$

if we adopt the first assumption ; or,

$$f = \frac{Pxy}{2nI} \tag{2}$$

if we adopt the second assumption.

Number of test.	Diam. of pulley.	Face.	Thickness of rim.	Width of hub.	Thickness of hub.	Length of arms.	Number of arms.	Dimensions of arms, all elliptic. at rim.	Dimensions of arms, all elliptic. at hub.	Breaking weight.	$f_1 = \frac{Pxy}{nI}$	$f_2 = \frac{Pxy}{nI}$	Place and Manner of Fracture.	Remarks.
1	14	4	$\frac{1}{1}$	$4\frac{1}{4}$	$\frac{3}{4}$	$4\frac{1}{4}$	5	$1\frac{1}{8} \times \frac{9}{16}$	$1\frac{7}{8} \times \frac{21}{32}$	5,600			Hub cracked.	Not a test of the arms.
2	15	$3\frac{1}{2}$	$\frac{5}{16}$	$3\frac{15}{16}$	$1\frac{3}{16}$	5	5	$1\frac{3}{8} \times \frac{11}{16}$	$1\frac{1}{2} \times \frac{13}{16}$	5,300			Hub cracked.	Not a test of the arms.
3	$12\frac{1}{2}$	$3\frac{3}{16}$	$\frac{5}{16}$	$3\frac{1}{4}$	$\frac{3}{4}$	$4\frac{1}{4}$	6	$2 \times \frac{7}{16}$	$2 \times \frac{7}{16}$	2,200	24,425	12,212	All the arms broke at the hub.	
4	$12\frac{1}{2}$	$3\frac{3}{16}$	$\frac{5}{16}$	$3\frac{1}{4}$	$\frac{3}{4}$	$4\frac{1}{4}$	6	$2 \times \frac{7}{16}$	$2 \times \frac{7}{16}$	2,100	23,314	11,657	All the arms broke at the hub.	
5	12	3	$\frac{1}{4}$	$3\frac{1}{8}$	$\frac{11}{16}$	$3\frac{1}{2}$	5	$1\frac{5}{16} \times \frac{1}{2}$	$1\frac{5}{8} \times \frac{9}{16}$	6,700	32,160	16,080	One arm broke at rim and hub.	Load subsequently increased to 8,000 when the rim broke.
6	$14\frac{3}{4}$	$3\frac{1}{4}$	$\frac{5}{16}$	$3\frac{3}{4}$	$\frac{11}{32}$	5	6	$1 \times \frac{9}{16}$	$1\frac{1}{4} \times \frac{5}{8}$	4,400	38,245	19,122	All the arms broke at the hub.	
7	15	$3\frac{7}{16}$	$\frac{1}{4}$	4	$\frac{9}{16}$	5	5	$1\frac{5}{16} \times \frac{3}{4}$	$1\frac{1}{2} \times \frac{27}{32}$	4,300	23,060	11,530	One arm broke at the hub.	Load subsequently increased to 5,300 when the rim broke.
8	24	$3\frac{1}{16}$	$\frac{5}{16}$	$3\frac{1}{4}$	$\frac{5}{8}$	$9\frac{5}{8}$	5	$1\frac{1}{8} \times \frac{11}{16}$	$1\frac{9}{16} \times \frac{3}{4}$	2,000	21,430	10,715	One arm broke at the hub.	Load subsequently increased to 2,200 when the rim broke.
9	14	4	$\frac{5}{16}$	$4\frac{1}{4}$	$1\frac{5}{32}$	4	5	$1\frac{1}{16} \times \frac{1}{2}$	$1\frac{7}{16} \times \frac{5}{8}$					One of the arms was broken in driving it out of the rim, so no test was made.
10	$13\frac{1}{2}$	$4\frac{1}{4}$	$\frac{5}{16}$	$4\frac{1}{4}$	$1\frac{1}{32}$	$4\frac{1}{2}$	5	$1\frac{7}{16} \times \frac{3}{4}$	$1\frac{11}{16} \times \frac{25}{32}$					There was a bushing inside the hub keyed to shaft, pulley slipped on bushing, hence no test.
11	15	$3\frac{7}{16}$	$\frac{5}{16}$	4	$\frac{9}{16}$	5	5	$1\frac{5}{16} \times \frac{3}{4}$	$1\frac{1}{2} \times \frac{27}{32}$	4,300	23,060	11,530	One arm broke at the hub during test of keys.	
12	$19\frac{3}{4}$	4	$\frac{1}{4}$	$4\frac{1}{4}$	$\frac{3}{4}$	$7\frac{1}{8}$	5	$1\frac{5}{8} \times \frac{25}{32}$	$1\frac{15}{16} \times \frac{25}{32}$					This pulley was the one used in testing keys and one of the arms broke during a key test.
Average.											26,528	13,264		

These formulæ are both based upon the assumption of arms of uniform section, either straight or else symmetrical with respect to hub and rim.

Other formulæ might be deduced which assume a variable section, but it would not seem to be worth while, in view of the fact that the bending moment is probably unequally divided among the arms. Hence the students confined themselves to computing the values of f from each of the above formulæ, thus obtaining average values of the constants to be used in these formulæ for the purpose of determining approximately the strength of the pulleys. (See table of the results on previous page.)

CONCLUSIONS FROM THESE TESTS.

1st. A low value of the modulus of rupture of cast iron should be used in the ordinary formulæ for designing pulley arms, due to the fact that a load at the rim acts more upon some arms than upon others, as shown by the fact that, in four out of eight of the tests, one arm broke first, and this one always occupied the same position.

2d In every case but one, of these four, a greater load than the original was afterwards put upon the pulley, and no other arm broke, but the rim gave way by crushing. In this one case excepted, the arms afterwards stood a greater load proportional to their number before breaking.

3d. In the tests on the single arms to be described next, the modulus of rupture rose as high as 55,000 lbs. in some cases, and in no case went below 35,000 lbs.

TESTS OF THE SEPARATE ARMS.

In the cases of numbers, 5, 7, 8, 9 and 10, some of the arms were not broken, the rims were now broken off, and the remaining arms were tested separately, the pull being exerted by a yoke hung over the end of the arm, the lower end being attached to the link of the machine.

The arms were always placed so that the direction of the pull was tangent to the curve of the rim at the end of the arm. The actual outside fibre stress at fracture was then determined by calculation from the experimental results, and is recorded in the following table:

Number of Arm.	Dimensions of section at fracture: all elliptical.	Bend of arm with or against load.	Actual outside fibre stress at fracture.	Average modulus of rupture for each pulley.
5 -- 1	$1\frac{9}{16} \times \frac{17}{32}$	against	44.396	44,396
7 -- 1	$1\frac{1}{2} \times \frac{7}{8}$	against	36,802	
7 -- 2	$1\frac{15}{32} \times \frac{7}{8}$	against	39,537	
7 -- 3	$1\frac{17}{32} \times \frac{7}{8}$	with	46.407	40,915
8 -- 1	$1\frac{15}{32} \times \frac{13}{16}$	against	35,508	
8 -- 2	$1\frac{11}{32} \times \frac{25}{32}$	against	36,091	
8 -- 3	$1\frac{11}{32} \times \frac{13}{16}$	with	39,939	
8 -- 4	$1\frac{13}{32} \times \frac{13}{16}$	with	42,469	38,500
9 -- 1	$1\frac{7}{16} \times \frac{5}{8}$	against	41,899	
9 -- 2	$1\frac{7}{16} \times \frac{21}{32}$	against	44,148	
9 -- 3	$1\frac{7}{16} \times \frac{5}{8}$	with	55,442	47,163
10 -- 1	$1\frac{3}{4} \times \frac{13}{16}$	against	54,743	
10 -- 2	$1^{15} \times \frac{13}{16}$	against	50.943	
10 -- 3	$1\frac{13}{16} \times \frac{13}{16}$	against	38,605	
10 -- 4	$1\frac{7}{8} \times \frac{13}{16}$	with	55,229	49.880

Total....... 642,153
Average..............................42,810

In order to show how the results in the preceding table were deduced from the experiments, the calculation will now be given in full for the first, or 5–1 (Fig. 33).

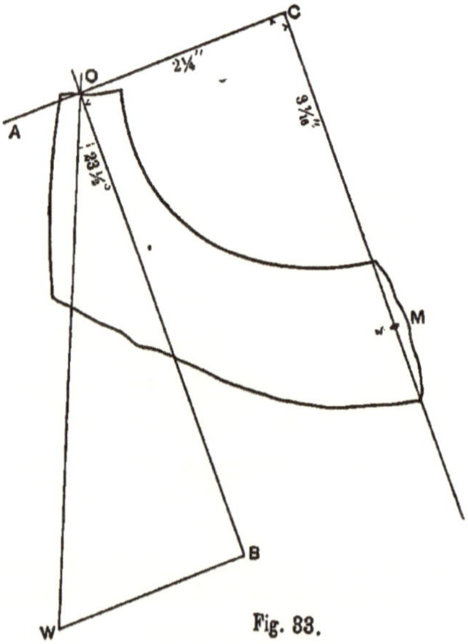

Fig. 33.

The force OW, which is equal to the load upon the arm, is resolved into two components, OB and BW. Both these components act on the arm at the point O, OB in the direction OB, and BW in the direction OA.

The first OB acts as a pull at the end of a cantilever of length OC, and is calculated accordingly; the second BW acts as a pull in the direction OA, and produces stresses similar to those acting in a hook, where the distance from the line of pull to the centre line of the hook tension is CM.

The formula used for the cantilever is $f_1 = \dfrac{My}{I}$, where M equals the pull times the length of the arm, y equals half the depth and I equals the moment of inertia of the section.

$$\frac{y}{I} = \frac{3r}{\pi b h^3}$$

The formula used to determine the greatest tension due to the force BW is

$$f_2 = \frac{P}{A} + \frac{Pny}{I}$$

where P equals the pull, A equals the area of the section $= \dfrac{\pi bh}{4}$, n equals the distance CM, and y equals the half depth.

The sum of f_1 and f gives us the greatest fiber stress at fracture, or the modulus of rupture of the iron of the arm. The breaking load of this arm was 1645 lbs. Hence:

$O W = 1645.$

$OB = O W \cos 23\tfrac{1}{2}° = 1508.$

$B W = O W \sin 23\tfrac{1}{2}° = 655.$

$$\therefore f_1 = \frac{(1508)\,(2\,25)\,(32)}{\pi\,(0.5312)\,(1.5625)^2} = 26671.$$

Also,

$$f_2 = \frac{(655)\,(4)}{\pi\,(0.5312)\,(1.562)} + \frac{(655)\,(3.437)\,(32)}{\pi\,(0\,5312)\,(1.562)^2} = 18725.$$

Hence $f_1 + f_2 = 44396$, as recorded in the table.

The other values are similarly calculated.

An inspection of the table will show that the modulus of rupture figures out higher when the bend of the arm is with the load than when it is against it, and the value will be found to be very much higher than the values of f derived for the pulleys with the rims on.

TESTS OF THE HOLDING POWER OF SET SCREWS.

These tests were all made by using pulley No. 12, the pulley being fastened to the shaft by two set screws and the shaft keyed to the holders; then the load required at the rim of the pulley to cause it to slip was determined, and this being multiplied by

$$\frac{9\frac{7}{8} + \frac{5}{16}}{\frac{27}{32} \times 2} = 6.037,$$

gives the holding power of the set screws.

The number 6.037 is obtained by adding to the radius of the pulley one-half the diameter of the wire rope, and dividing the sum by twice the radius of the shaft, since there were two set screws in action at a time. The set screws used were of wrought iron, $\frac{5}{8}$ of an inch in diameter, and having ten threads to the inch; the shaft used was of steel and rather hard, the set screws making but little impression upon it. The set screws were set up with a force of 75 lbs. at the end of a ten-inch monkey wrench. The set screws used were of four kinds, marked respectively A, B, C, and D. They may be described as follows:

A, ends perfectly flat, $\frac{9}{16}''$ diameter.

B, radius of rounded ends, about $\frac{1}{2}$ inch.

C, radius of rounded ends, about $\frac{1}{4}$ inch.

D, ends cup shaped and case hardened.

The results are given in the following table:

No. of test.	A	B	C	D
1	1,412	2,747	1,902	2,807
2	2,203	2,747	2,354	1,962
3	2,131	3,079	3,079	2,173
4	2,143	2,958	2,958	2,203
5	2,294	2,897		2,958
6	2,203	3,048		2,717
Av.	2,064	2,912	2,573	2,470

The following remarks should be made in regard to each kind of tests.

A. The set screws were not entirely normal to the shaft; hence they bore less in the earlier trials before they had become flattened by wear.

B. The ends of these set screws, after the first two trials, were found to be flattened, the flattened area having a diameter of about $\frac{1}{4}$ of an inch.

C. The ends were found, after the first two trials, to be flattened as in *B.*

D. The first test held well because the edges were sharp, then the holding power fell off till they had become flattened in a manner similar to *B*, when the holding power increased again.

KEYS.

The experiments on keys were made with pulley No. 11. In all cases where the keys were not as wide as the keyway they were wedged in with hardened steel pieces, the hardened steel piece in the pulley hub being as long as the hub was wide.

The load was applied as in the other tests, the shaft being firmly keyed to the holders. The load required at the rim of the pulley to shear the keys was determined, and this multiplied by a suitable constant, determined in a similar way to that used in the case of set screws, gives us the shearing strength per square inch of the keys.

The keys tested were of eight kinds, denoted, respectively, by the letters, *A*, *B*, *C*, *D*, *E*, *F*, *G* and *H*, and they may be described as follows :

A, were of Norway iron, $2'' \times \frac{1}{4}'' \times \frac{15}{32}''$; constant $= 18.5184$.

B, were of refined iron, $2'' \times \frac{1}{4}'' \times \frac{15}{32}''$; constant $= 18.5184$.

C, were of cast or tool steel, $1'' \times \frac{1}{4}'' \times \frac{15}{32}''$; constant $= 49.78$.

D, were of machinery steel, $2'' \times \frac{1}{4}'' \times \frac{15}{32}''$; constant $= 18.5184$.

E, were of Norway iron, $1\frac{1}{3}'' \times \frac{3}{8}'' \times \frac{7}{16}''$; constant $= 18.5184$.

F, were of cast iron, $2 \times \frac{1}{4} \times \frac{15}{32}$; constant $= 18.5184$.

G, were of cast iron, $1\frac{1}{3} \times \frac{3}{8} \times \frac{7}{16}$; constant $= 18.5184$.

H, were of cast iron, $1 \times \frac{1}{2} \times \frac{7}{15}$; constant $= 18.5184$.

The shearing stresses per square inch, as determined from the experiments, are given in the following table :

	A	B	C	D	E	F	G	H
1	41,202	36,482	100,056	70,186	37,036	34,166	38,700	29,814
2	41,759	37,334	91,344	66,110	37,222	36,944	37,222	38,978
3	40,184	39,254		64,630	36,850	30.278		
4	47,760	39,166		66,574		30,758		
Av.	42,726	38,059		66,875	37,036	33,034		

REMARKS.

A. Some crushing took place before shearing.

B. Slight crushing took place before shearing.

C. In the second test one of the wedges slipped and did not bear on the whole length of the key.

E. Inasmuch as these keys were only $1\frac{7}{8}''$ deep, they tipped slightly in the keyway.

H. In the first test there was a defect in the keyway of the pulley.

www.ingramcontent.com/pod-product-compliance
Lightning Source LLC
Chambersburg PA
CBHW021435090426
42739CB00009B/1484